MERCURY and VENUS

Robin Kerrod

Lerner Publications Company • Minneapolis

This edition published in 2000

Lerner Publications Company
A division of Lerner Publishing Group
241 First Avenue North, Minneapolis MN 55401 U.S.A.

Website address: www.lernerbooks.com

Library of Congress Cataloging-in-Publication Data

Kerrod, Robin.
    Mercury and Venus / Robin Kerrod.
        p. cm. – (Planet library)
    Includes index.
    Summary: Describes the physical features and
exploration of the two planets nearest to the Sun, Mercury,
and Venus.
    ISBN 0-8225-3904-7 (lib. bdg.)
    1. Mercury (Planet) Juvenile literature. 2. Venus (Planet)
Juvenile literature. [1. Mercury (Planet) 2. Venus
(Planet)] I. Title. II. Series: Kerrod, Robin. Planet library.
QB611.K47 2000   99-18352
523.41—dc21

Printed in Singapore by Star Standard Industries [PTE] Ltd
Bound in the United States of America
2 3 4 5 6 7 – OS – 07 06 05 04 03 02

CONTENTS

Introducing Mercury and Venus

Mercury and Venus are two of the planets in our solar system—the family of bodies that circle around the Sun. The solar system contains nine planets altogether, and Mercury and Venus are the only planets closer to the Sun than our home planet, Earth. Being so close to the Sun makes Mercury and Venus much hotter than Earth. In fact, temperatures on the two planets rise so high that they would melt metals such as tin and lead.

Like Earth, Mercury and Venus are made up mostly of rock. We call them terrestrial, or Earth-like, planets. Mars is the other terrestrial planet. But Mercury and Venus are quite different from Earth in most other ways. And they are also quite different from each other. For example, Venus is nearly the same size as Earth, but it is more than twice as big as Mercury. While Venus is surrounded by a thick atmosphere, or layer of gases, Mercury has only a small trace of an atmosphere.

Both Mercury and Venus have been known to astronomers for thousands of years. They can be easily seen with the naked eye, since they often shine brighter than the brightest stars. From our point of view on Earth, both planets stay quite close to the Sun. This means that we can see them only at sunrise or sunset.

Astronomers knew very little about Mercury and Venus until scientists began sending space probes to them. Even the most powerful telescopes on Earth show few features on Mercury's surface because the planet is so small and so far away. Venus is closer and larger, but we cannot see any of its surface from Earth because thick clouds always cover the planet.

Space probes, however, have taught us a great deal about the two planets. We know that Mercury is covered with craters and looks much like the Moon, while Venus is a land of huge volcanoes, unusual formations, and vast plains.

Mercury's surface is almost completely covered with craters. A space probe called *Mariner 10* photographed the planet in 1974.

Because Mercury and Venus circle closer to the Sun than Earth does, they always appear in the sky near the Sun. This means we can see them only just before sunrise in the east or just after sunset in the west.

Before sunrise, the sky is becoming lighter, but we can still see Mercury and Venus. They look like bright stars. At this time, we call them morning stars. Just after sunset, the sky is becoming darker, and Mercury and Venus again appear as bright stars. At this time, we call them evening stars.

SHOWING PHASES

As Mercury and Venus circle around the Sun, we see them change in size and shape. Their size appears to change because their distance from Earth is changing all the time. The closer they are to Earth, the bigger they appear in the sky.

The shapes of Mercury and Venus appear to change for another reason. At different times, we see more or less of the planets' surfaces lit up by the Sun. Like all the planets, Mercury and Venus give off no light of their own. They only reflect light from the Sun.

Sometimes just a small part of their surface is lit up. Then the planets appear as a thin crescent. At other times, we see half or more of the surface lit up. We call the changing shapes of the planets their phases. They are like the phases of the Moon, but we cannot see them with the naked eye. We can spot them only in telescopes.

orbit of Earth

orbit of Mercury

Mercury

Sun

Earth

Mercury circles the Sun inside Earth's orbit. So we see different parts of its surface lit up by the Sun at different times.

PLANETS IN TRANSIT

Once in a while, Mercury or Venus passes directly between Earth and the Sun in space. Then we see the planet pass over the Sun's surface. This is called a transit.

These times when a planet lines up between Earth and the Sun are rare. This is because the planets orbit, or circle, the Sun in different planes. Imagine the Sun on a sheet of paper, with Earth orbiting around it on the paper. Mercury and Venus do not tend to travel on the same imaginary piece of paper—they orbit the Sun in different planes. From Earth, they seem to travel slightly above or below the Sun.

Transits of Mercury happen about 15 times every century. They always occur around early May or mid November. Transits of Venus happen even less often, about twice every century.

The black dot on the Sun's surface is Mercury. It is making a transit of the Sun. Venus makes transits too, but not as often as Mercury.

Below: Earth orbits the Sun in one plane (top). Mercury and Venus orbit the Sun in a different plane (bottom).

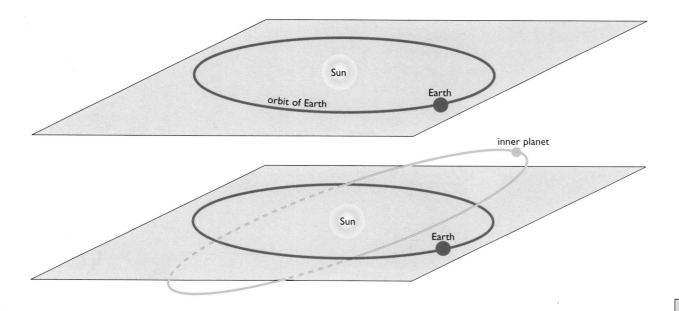

Planet Mercury

Mercury is a planet of many extremes. It is the planet closest to the Sun, the second smallest planet, and the fastest traveler around the Sun. The ancient Romans named it after the speedy messenger of their gods.

Mercury is a small planet. With a diameter of 3,031 miles (4,878 km), it is a little over a third the size of Earth.

The ancient Romans knew Mercury as both an evening star and a morning star. But they thought it was two different bodies. They called it Mercury as an evening star, and Apollo as a morning star.

RACING AROUND THE SUN

Mercury travels in its orbit, or path, around the Sun at an amazing speed of 107,000 miles (172,000 km) per hour. It takes the same amount of time as 88 days on Earth to orbit the Sun once. This is only one-fourth of the time Earth takes to travel around the Sun.

Mercury travels in an oval-shaped, or elliptical, orbit. At times it gets as close to the Sun as 28 million miles (46 million km). At other times, it wanders more than 43 million miles (60 million km) away.

Mercury is the nearest planet to the Sun, at an average distance of about 36 million miles (58 million km). It takes 88 Earth-days to circle the Sun once. Its immediate neighbor in space is Venus.

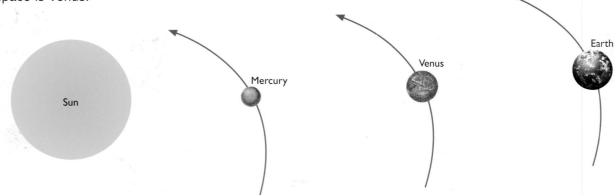

Sun

Mercury

Venus

Earth

Ancient symbol for
the planet Mercury

Mercury's barren
surface is covered
by thousands of
craters.

9

Mercury's
axis

Mercury's
orbit

Mercury spins around slowly on its axis, taking nearly two Earth-months to rotate once.

In a Spin

Like all the planets, Mercury moves in space in another way. It rotates, or spins around on its axis like a top.

A planet's axis is an imaginary line that runs through its center from its north pole to its south pole. Earth rotates on its axis once every 24 hours, or 1 day. Mercury spins around relatively slowly, taking nearly 59 Earth-days to rotate once.

This slow spin means that each "day" and "night" on Mercury are very long. Imagine that you travel to Mercury

Because of Mercury's slow rotation and its fast orbit, a day on Mercury—the time between sunrise and sunset—lasts 88 Earth-days.

and arrive at a place where the Sun is just rising. You will see the Sun climb very slowly into the sky. About 44 Earth-days will pass before it is "noon" on Mercury, with the Sun high overhead. And another 44 days will pass before the Sun sets. At this time, the planet will have traveled once around the Sun. Mercury's long day is then followed by an equally long night. The Sun does not rise again for another 88 days.

HOT AND COLD

Because of its long days and nights, Mercury is both a very hot and a very cold place. During the long day, the Sun beats down on part of the planet for nearly three Earth-months at a time. It bakes the planet's surface to 840° F (450° C), which is twice as hot as most home ovens.

After the Sun sets, Mercury's surface quickly cools down. Unlike Venus, Mercury has almost no atmosphere to act as a blanket to keep in the heat. All the heat escapes into space, and the temperature falls to about –290° F (–180° C). This is much, much colder than it has ever been on Earth.

If you lived on Mercury, you would see the Sun appear to change size during the day, as the planet moved closer or farther away from it.

840° F 450° C

Mercury is baking hot and freezing cold at the same time. This happens because it turns on its axis so slowly.

Sun

–290° F –180° C

MERCURY DATA

Diameter at equator: 3,031 miles (4,878 km)
Average distance from Sun: 36,000,000 miles (58,000,000 km)
Rotates on axis in: 58.7 Earth-days
Orbits Sun in: 88 Earth-days   **Moons:** None

Mercury is made up of three main parts—the crust, the mantle, and the core. The planet is not much bigger than the Moon, which is shown below for comparison.

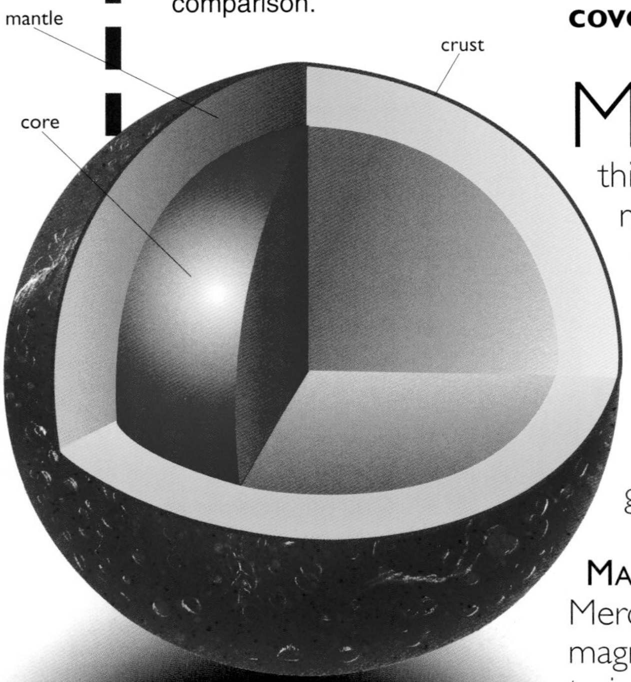

mantle

core

crust

Inside Mercury

Like Earth, Mercury is a rocky planet made up of different layers. Like Earth's Moon, Mercury is covered by thousands of craters.

Mercury has three main layers. Its outer layer, called the crust, is made up of hard rock. Beneath the crust is a thicker layer of rock known as the mantle. Beneath the mantle is a great ball of iron, which forms the planet's core. Mercury has a large amount of iron for its size. In fact, it has a greater percentage of iron than any other body in our solar system.

Surrounding the planet is a very thin atmosphere. It is only one-trillionth as thick as Earth's atmosphere. A few traces of gas are found near Mercury's surface. The gases present include helium, hydrogen, and sodium.

MAGNETIC MERCURY

Mercury's huge iron core is probably the reason for its magnetism. Magnetism is the force that draws certain metals to iron, and magnetic forces are created when masses of iron rotate. Earth also has an iron core, and the magnetism created is the force that makes a compass always point north. Earth's magnetism is stronger than Mercury's because our planet rotates much faster.

MERCURY'S SURFACE

Craters cover most of Mercury's surface. They were created by lumps of rock from outer space that have crashed down on the planet for billions of years.

Astronomers believe that most of the craters on Mercury were formed about 4 billion years ago, around 600 million years after the solar system formed. At the time, the solar system was full of rocky lumps, which rained down on planets and moons. Some were meteorites, or lumps of rock that range in size from tiny pebbles to huge chunks thousands of feet across. Others were larger lumps that measured miles across. These are known as the asteroids, which still circle the Sun in a large band between Mars and Jupiter.

When a lump of rock strikes a planet's surface, it creates a crater. The largest meteorites and asteroids made craters hundreds of miles across. These craters resemble the Moon's large craters, with raised rims and deep floors. They also have a small range of mountains in the middle.

When meteorites hit Mercury, masses of rocky fragments were thrown out. Large fragments fell back to the surface and made their own smaller craters. Particles of dust settled in spoke-like patterns around the craters. They reflect sunlight well and show up as shining crater rays.

STAR POINT

Mercury's faint atmosphere is less than one-trillionth as thick as Earth's atmosphere.

A close-up view of Mercury's surface. All the craters were made by lumps of rock from outer space that rained down on the planet billions of years ago.

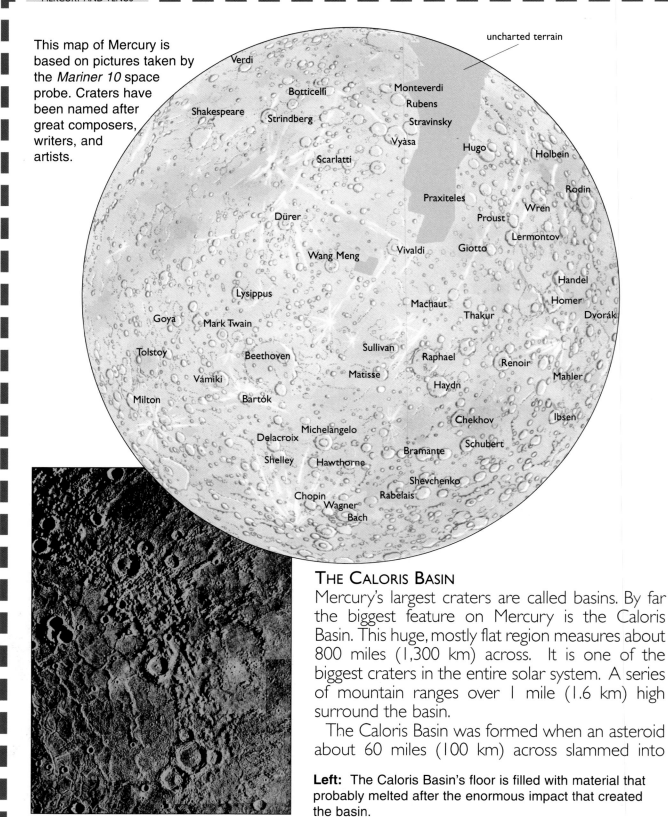

This map of Mercury is based on pictures taken by the *Mariner 10* space probe. Craters have been named after great composers, writers, and artists.

uncharted terrain

Verdi
Botticelli
Monteverdi
Shakespeare
Rubens
Strindberg
Stravinsky
Vyàsa
Hugo
Holbein
Scarlatti
Rodin
Praxiteles
Wren
Dürer
Proust
Lermontov
Wang Meng
Vivaldi
Giotto
Handel
Lysippus
Homer
Machaut
Goya
Thakur
Dvorák
Mark Twain
Tolstoy
Sullivan
Raphael
Renoir
Beethoven
Matisse
Mahler
Vàmìki
Haydn
Milton
Bartók
Chekhov
Ibsen
Michelangelo
Delacroix
Schubert
Shelley
Bramante
Hawthorne
Shevchenko
Chopin
Rabelais
Wagner
Bach

THE CALORIS BASIN

Mercury's largest craters are called basins. By far the biggest feature on Mercury is the Caloris Basin. This huge, mostly flat region measures about 800 miles (1,300 km) across. It is one of the biggest craters in the entire solar system. A series of mountain ranges over 1 mile (1.6 km) high surround the basin.

The Caloris Basin was formed when an asteroid about 60 miles (100 km) across slammed into

Left: The Caloris Basin's floor is filled with material that probably melted after the enormous impact that created the basin.

Mercury billions of years ago. The force of the impact made the planet's surface ripple, like water in a pond ripples when a stone is thrown into it. The ripples became the rings of mountains around the basin.

The impact of the asteroid also sent shock waves through the underground rocks. The waves traveled all the way to the other side of the planet, then shook and cracked the surface there. The result is a strange, bumpy landscape unlike any other part of Mercury. Astronomers call it the "weird terrain."

Ludwig van Beethoven (1770-1827)

What's in a Name?

Astronomers have named the various features on Mercury's surface after famous people and ships. Craters have been named after great composers (Beethoven and Vivaldi), artists (Michelangelo and Renoir), and writers (Tolstoy and Shelley).

Scarps were named after famous ships of exploration. Discovery Rupes was named after the ship that Captain James Cook sailed from Great Britain to the Hawaiian Islands in 1778.

PLAINS AND CLIFFS

Not all of Mercury is heavily cratered. About 60 percent of the planet's surface is covered by large flat regions, or plains. Also known as planitia, they are smooth like the flat plains on the Moon we call maria, or seas. They probably formed when lava, or hot liquid rock from underground, forced its way up through cracks in the surface. Then the pools of lava cooled and hardened to form the smooth plains.

In several places on Mercury there are steep cliffs known as scarps. Scarps are long, rounded cliffs that rise from 1,000 feet (300 m) to nearly 2 miles (3.2 km) high. Scarps stretch from 10 to 300 miles across Mercury's surface. No other planet has these formations, and neither does our Moon. One scarp called Discovery Rupes stretches for more than 300 miles (500 km) near the southern edge of the planet. Astronomers think that scarps are blocks of Mercury's crust that were forced upward when the planet cooled and shrunk after it formed.

STAR POINT

Beethoven is the largest crater on Mercury that we know of. It measures about 400 miles (650 km) across.

This picture shows one of the cliffs, or scarps, that are found on Mercury's surface. They can run for hundreds of miles.

Planet Venus

Venus is the planet that travels nearest to Earth. It shines brighter than any other object in the sky, except for the Sun and the Moon. The ancient Romans named Venus after their goddess of love and beauty.

Most people think of Venus as the evening star. We often see it shining brilliantly in the western sky just as the Sun goes down. We do not usually see it as a morning star because it appears so early in the morning, before sunrise.

VENUS IN MOTION

Venus travels in an almost perfect circle as it orbits the Sun at a distance of about 67 million miles (108 million km). Its path is only slightly elliptical. It takes about 225 Earth-days to travel once around the Sun.

Like all the planets, Venus also rotates on its axis. But it spins very slowly, taking 243 Earth-days to rotate once. So Venus takes longer to rotate once than it does to travel around the Sun. Venus's day (243 Earth-days) is longer than its year (225 Earth-days).

All the other planets rotate toward the east. On Earth, this means that the Sun appears to rise in the east and set in the west. However, Venus spins in the opposite direction, toward the west. So if you lived on Venus, you would see the Sun slowly rise in the west and set in the east.

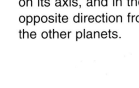

Venus's axis

Venus's orbit

Venus spins very slowly on its axis, and in the opposite direction from the other planets.

Venus is only a little smaller than Earth. Its diameter is about 400 miles (650 km) less than Earth's.

The thick clouds that cover Venus look white in ordinary photographs, but colored patterns show up in images sent back by space probes.

A Layered Planet

All we see of Venus from Earth is its cloudy atmosphere. But beneath the atmosphere is a rocky planet with a makeup similar to that of Earth. Like Earth and Mercury, Venus has different layers. Venus's atmosphere is so thick that it can be considered one of the planet's layers.

Underneath the atmosphere is the hard rocky outer layer called the crust. Scientists don't know its exact thickness, but it is probably 10–30 miles (15-50 km) deep. In the deepest part of the crust, the rock is probably so hot that it is molten, or liquid. Beneath the crust is a thick layer of heavier rock called the mantle. At Venus's center is a large metal core, made up mainly of iron and nickel. It may be partly molten, like Earth's metal core.

On Earth, the metal core produces Earth's magnetism. This happens because our planet rotates rapidly in space. Venus, however, rotates too slowly to create magnetism.

Right: Beneath its layer of clouds, Venus has a hard, rocky crust. Underneath is another rocky layer called the mantle, and beneath this a metal core.

STAR POINT

Venus is the hottest planet in our solar system. Temperatures at the surface rise as high as 900° F (480° C).

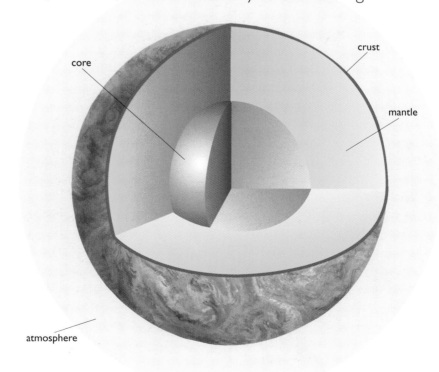

core

crust

mantle

atmosphere

Venus's Atmosphere

Venus's thick atmosphere has a crushing pressure. The clouds that cover the planet are made up of tiny drops of acid.

The atmosphere of Venus is quite different from the atmosphere we have on Earth. Earth's atmosphere is made up mainly of nitrogen and oxygen, the gas we must breathe to stay alive. Venus's atmosphere contains only a little nitrogen and no oxygen. The main gas is carbon dioxide, a very heavy gas. All that carbon dioxide makes the whole atmosphere heavy. Gravity causes it to press down on Venus's surface with more than 90 times the force of Earth's atmosphere. In other words, the atmospheric pressure on Venus is 90 times what it is on Earth.

UP IN THE CLOUDS

On Earth, the highest clouds form about 6 miles (10 km) above our planet's surface. Clouds in Venus's atmosphere extend much higher. Above Venus, the main cloud layers form at about 30 miles (50 km) above the planet's surface. The clouds extend up to about 65 miles (100 km) above the planet's surface.

Earth clouds are made up of tiny droplets of water or ice. On Venus, clouds are made up mainly of tiny drops of sulfuric acid. Sulfur is released into the atmosphere when volcanoes erupt on the surface. Chemical reactions between sulfur gas and water vapor (water in its gas form) change the sulfur into acid.

THE RUNAWAY GREENHOUSE

Besides providing oxygen for us to breathe, Earth's atmosphere also helps keep our

Several layers of cloud make up Venus's atmosphere. They occur at different levels and circulate in different directions.

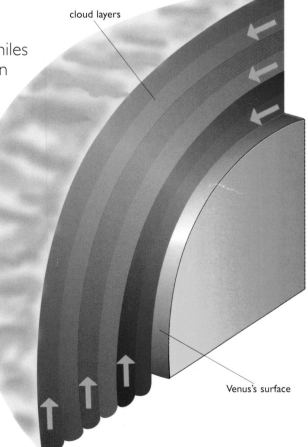

cloud layers

Venus's surface

planet warm. It acts like a greenhouse by letting in heat from the Sun and keeping some of it from escaping back into space. Many scientists believe that Earth's greenhouse effect is causing our average temperatures to rise. They think this is happening because larger amounts of heavy gases, such as carbon dioxide, are building up in our atmosphere and trapping more of the Sun's heat.

Venus's atmosphere also has a greenhouse effect. In fact, it has turned the planet into a world hotter than an oven. On Venus, the large amounts of carbon dioxide in the atmosphere trap most of the Sun's heat. This keeps the planet at temperatures as high as 900° F (480° C). The temperature does not vary much from place to place or from day to night because the heavy atmosphere acts as such a good blanket, or greenhouse. Some scientists believe that by learning more about the greenhouse effect on Venus we can better understand the same process on our own planet.

Right: The clouds on Venus reflect most of the Sun's light. Only a small amount gets through. But this heat gets trapped by the carbon dioxide in Venus's atmosphere.

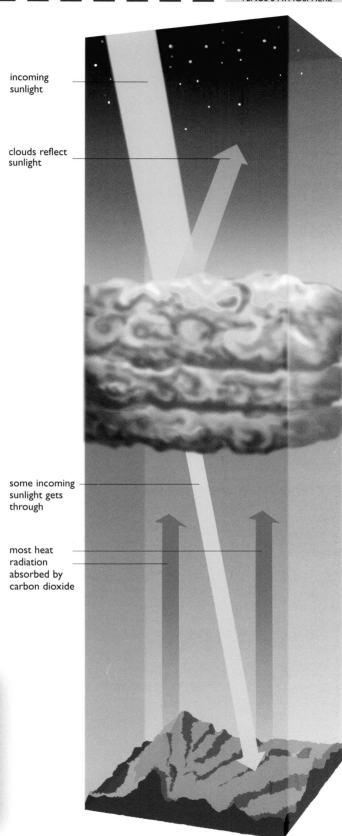

incoming sunlight

clouds reflect sunlight

some incoming sunlight gets through

most heat radiation absorbed by carbon dioxide

VENUS DATA

Diameter at equator: 7,521 miles (12,104 km)
Average distance from Sun: 67,200,000 miles (108,200,000 km)
Rotates on axis in: 243 Earth-days
Orbits Sun in: 224.7 Earth-days   **Moons:** None

Venus's Surface

Under its thick cloud covering, Venus has some amazing landscapes. There are a few highland areas, but most of the planet consists of vast rolling plains dotted with volcanoes.

The ancient symbol for the planet Venus

Venus may be similar to Earth in size, but its surface is quite different. More than two-thirds of Earth's surface is covered by the water of the oceans. Land areas, or continents, cover less than one-third of Earth's surface.

Venus has no great oceans or any water at all on the surface. The planet is too hot. If there ever were oceans on Venus, they would have boiled away a long time ago.

There are two large highland areas on Venus that we can think of as continents, like the continents on Earth. The largest one, Aphrodite Terra, lies close to Venus's equator. This continent is about the same size as South America. It features some spectacular volcanoes. One volcano, called Maat Mons, rises about 5.5 miles (9 km) high, about as tall as Earth's Mount Everest.

Volcanic mountains, rolling hills, and ancient lava flows are the main features of Venus's surface. This picture was produced by computer from radar images sent back by the *Magellan* space probe.

ISHTAR TERRA

Ishtar Terra, the other main continent on Venus, lies farther north. Two main features dominate Ishtar Terra, which is about the size of Australia. One is a mountain range known as Maxwell Montes. Some of the peaks in this range soar as high as 7.5 miles (12 km). Ishtar Terra's other main feature is Lakshmi Planum—a vast plateau, or high plain, surrounded by mountains.

Venus has several other smaller highland areas. Two of them are known as Alpha Regio and Beta Regio. Some highland areas are topped by ancient volcanoes.

This map of Venus was prepared from pictures taken by space probes that used radar to look through the planet's cloud covering.

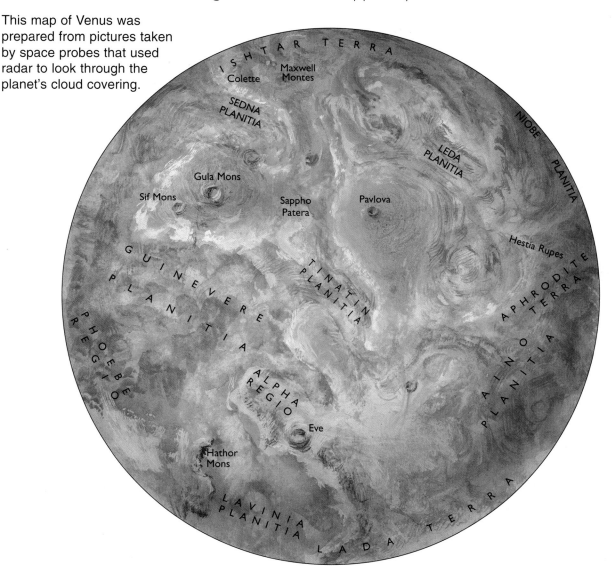

THE ROLLING PLAINS

Highland regions cover less than one-fifth of Venus's surface. The rest is covered by relatively flat plains. Most of these plains consist of gently rippling hills—we call them rolling plains. There are also lower and flatter plains regions, generally known as lowlands. Guinevere Planitia is one of the largest lowland regions, stretching for about 4,000 miles (7,000 km).

CRATERS ON VENUS

Like all the planets, Venus has been bombarded by lumps of rock called meteorites for billions of years. When meteorites struck the planet, they dug out craters in its surface. Venus has hundreds of craters, but nowhere near as many as can be found on Mercury.

The main reason for Venus's smaller number of craters is that it has a young surface. Great volcanic eruptions have changed the landscape in the past few hundred million years. In the time scale of our solar system, this is fairly recently. These "recent" lava flows covered up all the old craters. So the craters we see on Venus were created within the last few hundred million years. In contrast, most of the craters on Mercury are billions of years old.

There are few small craters on Venus's surface. Small meteorites burn up in Venus's thick atmosphere before they reach the planet's surface. So it is mostly larger meteorites that crash into Venus's surface.

Meteorites blast out masses of rocky surface material when they strike a planet. On Venus, some material settled around the crater, and some was blown away by wind. In places, we can see streaks formed by the wind-blown dust. Rippling dunes, like the sand dunes in deserts on Earth, are also visible.

Feminine Venus

The planet Venus was named after an ancient Roman goddess. Astronomers decided that many of the planet's features should be named after other goddesses or after famous women throughout history. Aphrodite Terra was named after the ancient Greek goddess of love. Guinevere Planitia was named for Queen Guinevere, wife of the legendary King Arthur. Nightingale Corona was named after Florence Nightingale, a famous English nurse. Crater Mead was named after Margaret Mead, the American anthropologist.

This view of Venus's surface was taken by the *Magellan* probe.

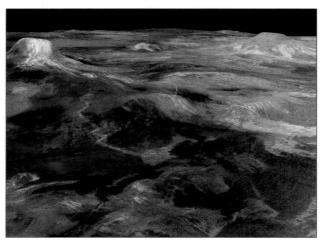

Shaping the Surface

Volcanoes have been the main force at work in shaping Venus's surface. Hundreds of them dot the landscape, and scientists believe many are still active.

On Earth, we know of three main kinds of volcanoes—cinder cones, shield volcanoes, and composite volcanoes. Cinder cones give off thick, sticky lava that does not flow very far. These volcanoes usually form a steep, cone-shaped mountain. Mount St. Helens in Washington State is a cinder cone.

Shield volcanoes give off very runny lava, which flows a long way. These volcanoes grow into a broad, flat mountain. The volcanoes of Hawaii are shield volcanoes. Composite volcanoes are a combination of the other two kinds. Mount Fuji in Japan is a composite volcano.

Venus's biggest volcanoes measure up to 300 miles (500 km) across and rise several miles high. But most are just a few miles across and a few hundred feet high.

Most of the volcanoes on Venus are shield volcanoes. Over time, the volcanoes poured out runny lava over and over again. This lava flowed widely. When it cooled and hardened, it formed the great rolling plains of Venus's surface.

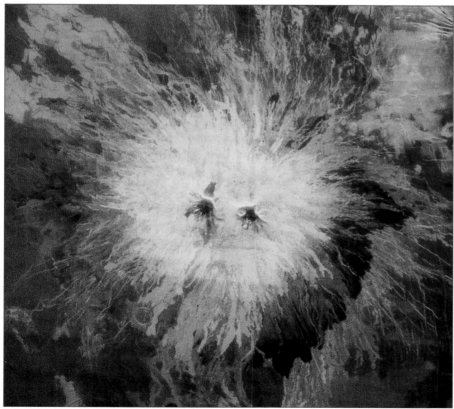

This is one of the many volcanoes found on Venus. It is at least 1 mile (1.6 km) high and measures more than 250 miles (400 km) across. It has erupted many times and may still be active.

Above: This is a close-up picture of a pancake dome, a feature not seen anywhere else in the solar system. It measures about 15 miles (25 km) across and is about 2,500 feet (800 m) high.

Below: This corona on Venus was produced by volcanic action. It measures about 250 miles (400 km) across.

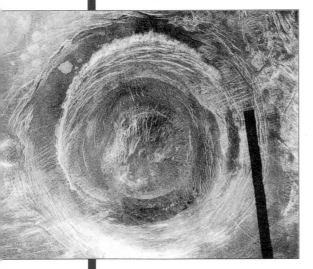

CHANNELS AND PANCAKES

The volcanoes of Venus have created many interesting features on the planet's surface. Flowing lava formed snaking channels that are hundreds of miles long. They look like dried-up riverbeds on Earth.

Eruptions have also created another interesting feature on Venus—the pancake domes. These flat, circular structures are usually tens of miles across and only a few thousand feet high. Nothing like them has been found anywhere else in the solar system.

CROWNS AND SPIDERS

Other features unique to Venus are the tall, circular coronae, which means crowns. Coronae are surrounded by a ring of ridges and troughs. Scientists believe coronae were formed by an upward movement of hot material from deep inside Venus. As the material cooled, the upper layers rose and fell, creating cracks in the surface.

Smaller than coronae but similar to them are Venus's arachnoids. These circular structures look much like spiderwebs, and the word *arachnoid* means like a spider. Arachnoids measure from 30 to 138 miles (50 to 230 km) across. They have a volcanic peak in the center, surrounded by a network of fine cracks. Arachnoids may have been created by molten lava pushing up from below.

The Tortured Crust

The fine cracks, or fractures, that cover Venus's coronae and arachnoids were formed when the rocks in the crust moved and split. More widespread movements in Venus's crust have created other surface features, including mountain ranges. Movements of rock up or down along faults, or weaknesses, in the crust have created long troughs and ridges. Some of these troughs and ridges run for thousands of miles.

Even more prominent are the valleys, or rifts, that follow fault lines in many places. Wide rift valleys are found both in the plains and in the highland regions. The continent Aphrodite Terra is riddled with rift valleys. One called Diana Chasma is over 250 miles (400 km) wide in places.

Strange features called tesserae were also caused by movements in Venus's crust. These criss-cross patterns of ridges and grooves appear on no other body in the solar system.

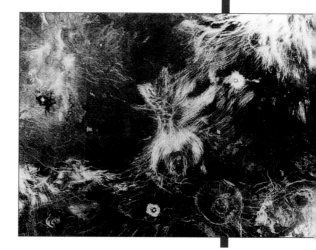

Above: This region of Venus's surface is about 1,000 miles (1,600 km) wide. It shows a number of the spidery features called arachnoids.

An artist's impression shows the highest region on Venus, Maxwell Montes

Probing Mercury and Venus

Astronomers knew very little about what Mercury and Venus were like until they sent space probes to them.

All the close-up pictures we have of Mercury were taken by the NASA (National Aeronautics and Space Administration) space probe *Mariner 10*, which visited the planet in 1974.

Mariner 10 set out for Mercury in November 1973, launched from Cape Canaveral, Florida. But it did not fly to Mercury by the most direct route. First it was sent to fly by Venus. It did this to pick up speed. It used the pull of Venus's gravity to make it move faster.

It took nearly three months for *Mariner 10* to reach Venus, where it took over 4,000 pictures of the cloud-covered planet in February 1974. Venus's gravity speeded it up and boosted it into a trajectory (path) that took it to Mercury in March. It flew as close as 435 miles (700 km), taking pictures that showed how much Mercury looked like the Moon.

After leaving Mercury, *Mariner 10* looped around the Sun before flying past Mercury once more in September, though much farther away (more than 25,000 miles, or 40,000 km). Six months later, it returned again, this time skimming only about 200 miles (300 km) above the planet's surface.

The space probe *Mariner 10* had twin television cameras that took pictures of Mercury and Venus. Picture signals were transmitted by the dish antenna. Twin solar panels, each nearly 9 feet (2.7 m) long, supplied power to the craft. A sun shade protected the main body of the spacecraft from the Sun's heat.

American scientists launched the first successful probe to Venus in 1962. *Mariner 2* sent back the first reports of Venus's very high temperature. Russian scientists achieved the next success when they parachuted instruments from their probe *Venera 4* into Venus's atmosphere in 1967. It confirmed that the planet was very hot and reported that it had a thick atmosphere of mostly carbon dioxide.

RUSSIAN LANDINGS

Three years later, the Russian probe *Venera 7* actually landed on Venus's surface. It sent back information about Venus's atmosphere, but no pictures. In 1975, however, *Venera 9* and *Venera 10* landed and sent back pictures. The pictures showed a number of what looked like volcanic rocks.

By this time, the American probe *Mariner 10* had photographed Venus from space on its way to Mercury in 1974. At its closest, the probe swooped to within 4,000 miles (6,400 km) of the planet's surface. Its photographs showed bands of clouds swirling in the thick atmosphere.

Above: This Russian *Venera* probe explored Venus in the 1970s. The ball-shaped capsule at the bottom of the craft was dropped into the atmosphere and floated down to the surface by parachute.

Right: *Mariner 10* took this picture of Venus in ultraviolet light. In this kind of light, the bands of clouds swirling about in the planet's atmosphere show up clearly.

RADAR IMAGES

While space scientists were sending probes to Venus, astronomers began using other ways to look at the planet. Astronomers used a system called radar (radio detecting and ranging) to send radio waves to Venus. The radio waves were reflected back and then collected and used to create radar images of the planet. These images showed general features of the planet's surface. Highland areas and fault lines were visible.

The *Pioneer Venus 1* probe was the first to use radar to look at Venus's surface. Launched in May 1978, it went into orbit around Venus seven months later.

SCANNING FROM ORBIT

The next leap forward in our knowledge about Venus came in December 1978, when *Pioneer Venus 1* began orbiting Venus. It sent back radar images of the planet, produced a map of its surface, and measured temperatures in its upper atmosphere. It discovered the main features of the landscape, such as the two main continents and the rolling plains.

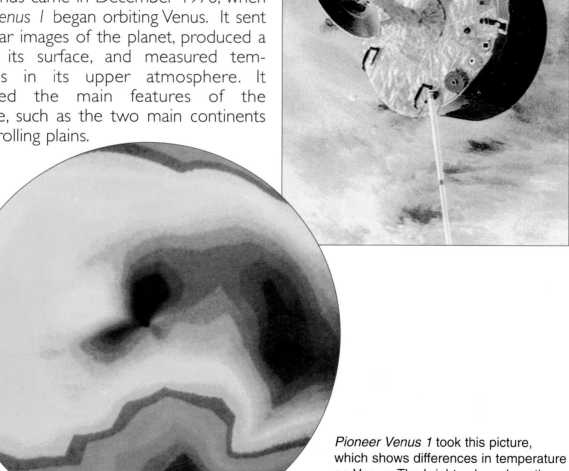

Pioneer Venus 1 took this picture, which shows differences in temperature on Venus. The bright colors show the hotter side lit by the Sun. The blue regions show lower temperatures on the dark side of the planet.

Radar Eyes

The word *radar* stands for radio detecting and ranging. Radar is used on Earth mainly to locate aircraft in the sky and ships at sea.

At an airport, for example, a beam of radio waves is sent from an antenna. When the beam hits an aircraft, it is reflected back. The antenna picks up the reflected beam and sends signals to a radar screen, like a TV screen. The signals make a dot, or blip, appear on the screen, which shows exactly where the aircraft is.

outgoing beam

reflected beam

antenna

Below: The *Magellan* probe orbited Venus for more than four years. It weighed over 3½ tons and had a dish antenna 12 feet (3.7 m) across. Its twin solar panels measured about 11 feet (3.5 m) across.

MARVELOUS *MAGELLAN*

The *Pioneer Venus* probe showed us what a fascinating place Venus is. But its instruments could not make out much detail. So space scientists decided to send a more advanced probe to the planet. *Magellan* was launched in May 1989. After 15 months, it reached its target, then it went into orbit around Venus in August 1990. Radar images from *Magellan* provided scientists with details about Venus's surface features. *Magellan* operated successfully for about four years, before falling into Venus's atmosphere in October 1994 and burning up like a shooting star.

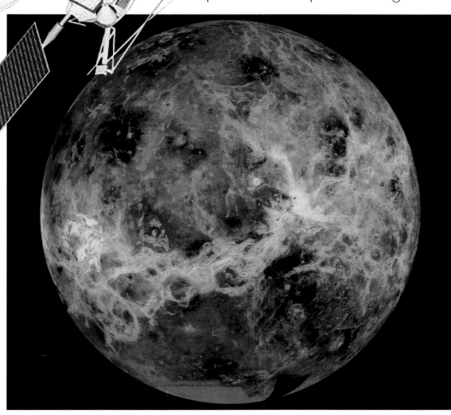

Left: The *Magellan* probe showed Venus to be unlike any other body in the solar system.

Life on Venus?

In the early 1900s, some people thought that Venus might be quite a pleasant world, something like the steamy jungles we have on Earth.

But we have learned that humans could not survive on Venus. If you landed on its surface, you would quickly be roasted by the high temperature. You would be crushed by the pressure. And you would be burned by a mist of sulfuric acid.

Some scientists believe Venus could be made into a world like Earth. They call this process terraforming, after the word *terra*, which means Earth. The first step would be to send a fleet of spacecraft to Venus carrying certain kinds of tiny plants called algae. Plants take in carbon dioxide and give off oxygen, so these plants could use up much of the carbon dioxide in Venus's atmosphere and give out oxygen.

As the plants used up the carbon dioxide, more heat could escape into space. Then the planet could cool down. If it became cool enough, rain could fall out of the clouds to form oceans. Then one day, the clouds could roll away, and the Sun would shine on Venus for the first time. Venus would be on the way to becoming another home for plants, animals, and people.

Some scientists also believe that undertaking a process like this could have another benefit. Because Venus's atmosphere is like an extreme version of the greenhouse effect we have on Earth, learning how to decrease the amount of carbon dioxide on Venus could help us learn how to do the same at home.

Some scientists once believed that conditions on Venus might be something like those on Earth millions of years ago. They thought Venus might have huge swampy forests of trees and giant ferns.

Glossary

arachnoid: a spiderweb-like feature on Venus's surface

asteroid: a type of rocky body that orbits the Sun between Mars and Jupiter

atmosphere: the layer of gases around a heavenly body

axis: an imaginary line running through a planet from its north to its south pole

core: the center part of a planet or moon

corona: a circular feature on Venus's surface

crater: a pit in the surface of a planet or a moon.

crust: the hard outer layer of a rocky planet like Mercury and Venus

evening star: the planet Mercury or Venus shining in the western sky just after sunset

gravity: the attraction, or pull, that every heavenly body has on objects on near it

greenhouse effect: when a planet's atmosphere acts like a solar greenhouse, trapping the Sun's heat on the planet

mantle: the layer of rock underneath the crust of a rocky planet

meteorite: a lump of rock or metal from outer space that falls to the surface of a planet or moon

morning star: the planet Mercury or Venus shining in the eastern sky just before sunrise

orbit: the path in space of one heavenly body around another, such as Mercury around the Sun

phases: the different shapes of Mercury, Venus, or the Moon, as seen from Earth, as more or less of them is lit up by the Sun

planet: a large body that circles in space around the Sun

probe: a spacecraft that travels from Earth to explore bodies in the solar system

scarp: a steep cliff

solar system: the Sun and all the bodies that travel around it, including Mercury and Venus

terraforming: making another world into a world more like Earth

terrestrial: like Earth

transit: the time when Mercury or Venus travels across the face of the Sun, as viewed from Earth

volcano: a place where molten (liquid) rock escapes from beneath the surface of a planet or moon

Index

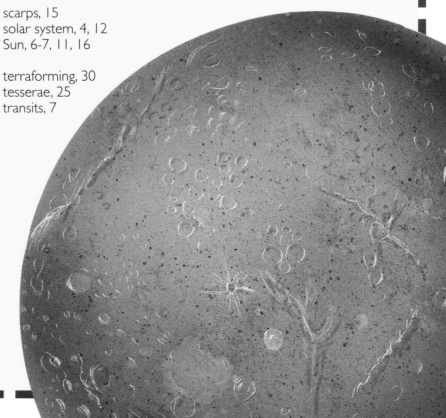